Renate & Uwe H. Sueltz

GAS STATION

BoD - Books on Demand
Norderstedt - Germany 2019

Bibliografische Information durch die Deutsche Nationalbibliothek
Die Deutsche Nationalbibliothek verzeichnet diese Publikation in der
Deutschen Nationalbibliografie; detaillierte bibliografische Daten
sind im Internet über http://dnb.dnb.de abrufbar.

© Renate & Uwe H. Sültz
Herstellung und Verlag:
BoD – Books on Demand, Norderstedt
ISBN 9-78374-1-29069-5

DATE 	TOTAL MILES ... DRIVEN 		GALLONS	PRICE PER GALLON $	GAS REGULAR PLUS PREMIUM DIESEL	AIR PRESSURE PSI
					drive carefully	

DATE	TOTAL MILES ... DRIVEN	GALLONS	PRICE PER GALLON $	GAS REGULAR PLUS PREMIUM DIESEL	AIR PRESSURE PSI

drive carefully

DATE	TOTAL MILES ... DRIVEN	GALLONS	PRICE PER GALLON $	GAS REGULAR PLUS PREMIUM DIESEL	AIR PRESSURE PSI
				drive carefully	

DATE	TOTAL MILES ... DRIVEN	GALLONS	PRICE PER GALLON $	GAS REGULAR PLUS PREMIUM DIESEL	AIR PRESSURE PSI
				drive carefully	

DATE	TOTAL MILES ... DRIVEN	GALLONS	PRICE PER GALLON $	GAS REGULAR PLUS PREMIUM DIESEL	AIR PRESSURE PSI
				drive carefully	

DATE 	TOTAL MILES ... DRIVEN	GALLONS	PRICE PER GALLON $	GAS REGULAR PLUS PREMIUM DIESEL	AIR PRESSURE PSI 
				drive carefully	

DATE	TOTAL MILES	... DRIVEN	GALLONS	PRICE PER GALLON	GAS REGULAR PLUS PREMIUM DIESEL	AIR PRESSURE PSI
					drive carefully	

DATE	TOTAL MILES ... DRIVEN	GALLONS	PRICE PER GALLON $	GAS REGULAR PLUS PREMIUM DIESEL	AIR PRESSURE PSI
				drive carefully	

DATE	TOTAL MILES ... DRIVEN	GALLONS	PRICE PER GALLON $	GAS REGULAR PLUS PREMIUM DIESEL	AIR PRESSURE PSI
				drive carefully	

DATE	TOTAL MILES ... DRIVEN	GALLONS	PRICE PER GALLON $	GAS REGULAR PLUS PREMIUM DIESEL	AIR PRESSURE PSI

drive carefully

DATE	TOTAL MILES	... DRIVEN	GALLONS	PRICE PER GALLON $	GAS REGULAR PLUS PREMIUM DIESEL	AIR PRESSURE PSI

drive carefully

DATE	TOTAL MILES / ... DRIVEN	GALLONS	PRICE PER GALLON ($)	GAS (REGULAR / PLUS / PREMIUM / DIESEL)	AIR PRESSURE PSI

drive carefully

DATE 	TOTAL MILES ... DRIVEN	GALLONS	PRICE PER GALLON $	GAS REGULAR PLUS PREMIUM DIESEL	AIR PRESSURE PSI 
				drive carefully	

DATE	TOTAL MILES ... DRIVEN	GALLONS	PRICE PER GALLON $	GAS REGULAR PLUS PREMIUM DIESEL	AIR PRESSURE PSI
				drive carefully	

DATE 	TOTAL MILES ... DRIVEN	GALLONS	PRICE PER GALLON $	GAS REGULAR PLUS PREMIUM DIESEL	AIR PRESSURE PSI 
				drive carefully	

DATE	TOTAL MILES / DRIVEN	GALLONS	PRICE PER GALLON $	GAS REGULAR / PLUS / PREMIUM / DIESEL	AIR PRESSURE PSI
				drive carefully	

DATE 	TOTAL MILES ... DRIVEN 	GALLONS 	PRICE PER GALLON $	GAS REGULAR PLUS PREMIUM DIESEL	AIR PRESSURE PSI 
				drive carefully	

DATE	TOTAL MILES ... DRIVEN	GALLONS	PRICE PER GALLON $	GAS REGULAR PLUS PREMIUM DIESEL	AIR PRESSURE PSI 

drive carefully

DATE 	TOTAL MILES ... DRIVEN	GALLONS	PRICE PER GALLON $	GAS REGULAR PLUS PREMIUM DIESEL	AIR PRESSURE PSI 

drive carefully

DATE	TOTAL MILES ... DRIVEN	GALLONS	PRICE PER GALLON $	GAS REGULAR PLUS PREMIUM DIESEL	AIR PRESSURE PSI

drive carefully

DATE	TOTAL MILES ... DRIVEN	GALLONS	PRICE PER GALLON $	GAS REGULAR PLUS PREMIUM DIESEL	AIR PRESSURE PSI
				drive carefully	

DATE 	TOTAL MILES ... DRIVEN	GALLONS	PRICE PER GALLON $	GAS REGULAR PLUS PREMIUM DIESEL	AIR PRESSURE PSI 

drive carefully

DATE	TOTAL MILES → ... DRIVEN ↓	GALLONS	PRICE PER GALLON $	GAS REGULAR PLUS PREMIUM DIESEL	AIR PRESSURE PSI

drive carefully

DATE	TOTAL MILES ... DRIVEN	GALLONS	PRICE PER GALLON $	GAS REGULAR PLUS PREMIUM DIESEL	AIR PRESSURE PSI
				drive carefully	

DATE 	TOTAL MILES ... DRIVEN	GALLONS	PRICE PER GALLON $	GAS REGULAR PLUS PREMIUM DIESEL	AIR PRESSURE PSI 

drive carefully

DATE	TOTAL MILES ... DRIVEN	GALLONS	PRICE PER GALLON $	GAS REGULAR PLUS PREMIUM DIESEL	AIR PRESSURE PSI

drive carefully

DATE 	TOTAL MILES ... DRIVEN	GALLONS	PRICE PER GALLON $	GAS REGULAR PLUS PREMIUM DIESEL	AIR PRESSURE PSI 
				drive carefully	

DATE	TOTAL MILES ... DRIVEN	GALLONS	PRICE PER GALLON $	GAS REGULAR PLUS PREMIUM DIESEL	AIR PRESSURE PSI
				drive carefully	

DATE	TOTAL MILES ... DRIVEN	GALLONS	PRICE PER GALLON $	GAS REGULAR PLUS PREMIUM DIESEL	AIR PRESSURE PSI
				drive carefully	

DATE	TOTAL MILES ... DRIVEN	GALLONS	PRICE PER GALLON	GAS REGULAR PLUS PREMIUM DIESEL	AIR PRESSURE PSI
				drive carefully	

DATE	TOTAL MILES ... DRIVEN	GALLONS	PRICE PER GALLON $	GAS REGULAR PLUS PREMIUM DIESEL	AIR PRESSURE PSI
				drive carefully	

DATE	TOTAL MILES ... DRIVEN	GALLONS	PRICE PER GALLON $	GAS REGULAR PLUS PREMIUM DIESEL	AIR PRESSURE PSI

drive carefully

AIR PRESSURE PSI 	GAS REGULAR PLUS PREMIUM DIESEL	PRICE PER GALLON $	GALLONS	TOTAL MILES ... DRIVEN	DATE 
	drive carefully				

DATE	TOTAL MILES / ... DRIVEN	GALLONS	PRICE PER GALLON $	GAS REGULAR PLUS PREMIUM DIESEL	AIR PRESSURE PSI
				drive carefully	

DATE 	TOTAL MILES ... DRIVEN	GALLONS	PRICE PER GALLON $	GAS REGULAR PLUS PREMIUM DIESEL	AIR PRESSURE PSI 
				drive carefully	

DATE	TOTAL MILES ... DRIVEN	GALLONS	PRICE PER GALLON	GAS REGULAR PLUS PREMIUM DIESEL	AIR PRESSURE PSI
				drive carefully	

DATE 	TOTAL MILES ... DRIVEN	GALLONS	PRICE PER GALLON $	GAS REGULAR PLUS PREMIUM DIESEL	AIR PRESSURE PSI 

drive carefully

DATE 	TOTAL MILES ... DRIVEN 	GALLONS	PRICE PER GALLON $	GAS REGULAR PLUS PREMIUM DIESEL	AIR PRESSURE PSI 
				drive carefully	

DATE	TOTAL MILES ... DRIVEN	GALLONS	PRICE PER GALLON $	GAS REGULAR PLUS PREMIUM DIESEL	AIR PRESSURE PSI
				drive carefully	

DATE	TOTAL MILES ... DRIVEN	GALLONS	PRICE PER GALLON $	GAS REGULAR PLUS PREMIUM DIESEL	AIR PRESSURE PSI
				drive carefully	

DATE	TOTAL MILES ... DRIVEN	GALLONS	PRICE PER GALLON $	GAS REGULAR PLUS PREMIUM DIESEL	AIR PRESSURE PSI

drive carefully

DATE	TOTAL MILES ... DRIVEN	GALLONS	PRICE PER GALLON $	GAS REGULAR PLUS PREMIUM DIESEL	AIR PRESSURE PSI

drive carefully

DATE	TOTAL MILES	... DRIVEN	GALLONS	PRICE PER GALLON $	GAS REGULAR PLUS PREMIUM DIESEL	AIR PRESSURE PSI
					drive carefully	

DATE	TOTAL MILES ... DRIVEN	GALLONS	PRICE PER GALLON $	GAS REGULAR PLUS PREMIUM DIESEL	AIR PRESSURE PSI
				drive carefully	

DATE	TOTAL MILES ... DRIVEN	GALLONS	PRICE PER GALLON $	GAS REGULAR PLUS PREMIUM DIESEL	AIR PRESSURE PSI
				drive carefully	

DATE	TOTAL MILES ... DRIVEN	GALLONS	PRICE PER GALLON $	GAS REGULAR PLUS PREMIUM DIESEL	AIR PRESSURE PSI

drive carefully

DATE	TOTAL MILES ... DRIVEN	GALLONS	PRICE PER GALLON $	GAS REGULAR PLUS PREMIUM DIESEL	AIR PRESSURE PSI

drive carefully

DATE	TOTAL MILES ... DRIVEN	GALLONS	PRICE PER GALLON	GAS REGULAR PLUS PREMIUM DIESEL	AIR PRESSURE PSI
				drive carefully	

DATE	TOTAL MILES	... DRIVEN	GALLONS	PRICE PER GALLON $	GAS REGULAR PLUS PREMIUM DIESEL	AIR PRESSURE PSI
					drive carefully	

DATE	TOTAL MILES ... DRIVEN	GALLONS	PRICE PER GALLON $	GAS REGULAR PLUS PREMIUM DIESEL	AIR PRESSURE PSI

drive carefully

www.ingramcontent.com/pod-product-compliance
Lightning Source LLC
Chambersburg PA
CBHW050025230526
45470CB00003B/1131